Nathaniel N Pierce

Directions for cleansing the blood and curing all forms of

disease that the human family is subject to

without the use of any medicine whatever

Nathaniel N Pierce

Directions for cleansing the blood and curing all forms of disease that the human family is subject to
without the use of any medicine whatever

ISBN/EAN: 9783744738194

Printed in Europe, USA, Canada, Australia, Japan

Cover: Foto ©berggeist007 / pixelio.de

More available books at **www.hansebooks.com**

THE BOOK OF BOOKS.

DIRECTIONS

FOR

CLEANSING THE BLOOD

*AND CURING ALL FORMS OF DISEASE
THAT THE HUMAN FAMILY
IS SUBJECT TO,*

WITHOUT THE USE OF ANY MEDICINE

WHATEVER.

By PROF. NATHANIEL N. PIERCE,

Psychometric Counselor and Healing Medium, of Philmont, N.Y.
formerly of Putnam, Conn.

[Copyrighted by N. N. Pierce, A. D., 1883.]

TENTH EDITION.

PRICE   CENTS.

NEW YORK:
PRINTED AT THE OFFICE OF THE TRUTH SEEKER.

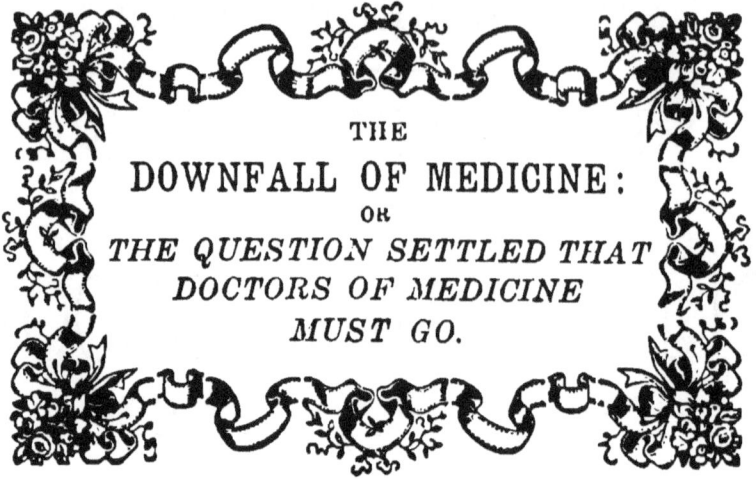

THE
DOWNFALL OF MEDICINE:

OR

THE QUESTION SETTLED THAT DOCTORS OF MEDICINE MUST GO.

CONTENTS.

CONTENTS.

CONTENTS.

PREFACE.

As a friend to suffering humanity, I have been impelled to write this book, "The Downfall of Medicine; or, The Question Settled." Doctors who are not in possession of sufficient intelligence to cure disease without dosing their patients with medicine, must go, and immediately after them must follow the entire destruction of the business of the druggist; the miserable pill makers; the patent medicine venders; Indian doctors and venders of roots and herbs, who have· encompassed the earth and the depths of the seas to find the most poisonous and loathsome plants, and the bodies and entrails of filthy insects and animals, each of which in turn has been made to contribute to dose the stomachs and bodies of poor suffering men, women and children.

But thanks to the Power that ever watches over the suffering children of men, the day of their relief has already dawned; the mysterious hand-writing has appeared upon the wall, the doom of medicines and vile drugs has been written in one magic word—WATER! bright, sparkling, and beautiful forever. When used as I have herein given instructions, will

cure every form of disease that the human
body is heir to; and the people of this country will become more intelligent, and will no
longer bow down to medicine mongers, but
will cure themselves by the divine curative
properties of water. Although roots, herbs,
drugs, and minerals may give temporary relief,
in nine cases out of ten they produce other
diseases, often much worse than the diseases
they are expected to cure: consequently, their
use as medicine must be abolished forever, and
water, bright and sparkling, must take their
places as a safe substitute for the cure of
every form of disease, external or internal,
that the bodies of men or beasts are subject
to.

THE AUTHOR.

Directions for Cleansing the Blood.

After an experience of thirteen years in healing all forms of disease without the use of any medicine, I have concluded to give to the world the benefit of that experience, by giving full and explicit directions for treating each and every form of disease. But before I do this, I will tell you, my friend, a few facts in regard to the present system or practice of medicine. In no country can you find so much imposition practiced as there is in America. Step by step have the doctors of this country been driven out of their cruel and barbarous practices, by the intelligence of the people, and the day is not far distant when all doctors' prescriptions to druggists will have to be given in plain English instead of Latin, and made compulsory by both custom and law; and when the study of medicine will be taught in the common schools, the same as any other branch of education taught; and instead of the heavy fees now exacted by the average doctor, his daily pay or wages must come down to an average of the common mechanic, who is the most useful man of the

two, in the community where he resides. I
will say further, that there is nothing in the
world that can excel the cheek and impudence
of the average doctor of the present day, unless
it is the silly, ignorant State legislators of a
few of these United States, who have been so
stupid as to pass laws to restrict and regulate
the practice of medicine, just because these
learned ignoramuses—doctors—asked them to
pass laws making it a crime punishable by fine
and imprisonment, even for some old grand-
mother to give one of her family a dose of cat-
nip tea, as then she would be practicing medicine contrary to these special laws passed by a
few states for the regulation of the practice of
medicine. Nothing in the history of our country has ever disgraced the statute books of any
state more than these same disgraceful doctors' laws, unless we refer back to the old Puri-
tan Blue Laws of Connecticut, my own native
state; but I am happy to say that this old
state has outgrown her bigotry, and for intelligence stands ahead of any state in the Union.

And here let me say a few words in regard
to quack medicines usually sold at the drug
stores throughout the land. When you see
some man's or woman's homely profile stuck
on the fences along the road-sides, and see
their homely portraits in nearly every news-
paper you chance to read, make up your mind
that *there* is a genuine fraud, for when anyone
finds a remedy truly beneficial to the human
family, people will soon find it out without

resorting to advertisements prefaced with homely portraits. Such advertisements serve only to catch gulls, at least you will find yourself gulled, my friend, if you chance to buy the vile nostrums.

Another class of quacks advertise under the head of "Indian Medicines" and "Indian Doctors," but don't be deceived by them; remember that the white man taught the Indian all that he ever knew about medicine, roots and plants, or anything else. Don't put any confidence in any knowledge purporting to come through any Indian source, for all this first came through the minds of white men, and you may rest assured that there has been no improvement in the remedy after having been distilled through the mind of an Indian. Bear in mind that the less medicine you take the longer you will live and the better health you will enjoy. Although some medicine may give you temporary relief, it makes you a slave for life to the habit of taking medicine.

I also wish to impress upon your mind the fact that all strong drinks are irritants. Use no beer, wine, or any drink that contains alcohol, except in case of the bite of some poisonous reptile, or for blood-poisoning caused by handling such reptiles—usually called poison by inoculation; in such cases one poison counteracts the other, and thus can be made beneficial. Otherwise don't allow anything that can intoxicate to be used outside the body, or inside; although it may give temporary relief, it leaves a poison behind that is taken up by the blood and deposited in the weakest parts of the body.

Again, don't employ any doctor who uses alcoholic drinks in his practice, and discharge any physician who may attempt malpractice on your person by the use of alcohol. Avoid as you would a pestilence all quack medicines. No wonder that many of these quack medicine venders, who have become immensely rich through the sale of their vile nostrums, have ended their days in insane asylums, for their surprise at unexpected success unbalanced their weak minds. I am happy to say that the patrons of these quacks are becoming less every day. As the people of this great Union become more enlightened, they are not so easily caught by the flaming advertisements of quack medicine venders, even though said quacks hail from Lowell, Lynn, or Roxbury, Mass., or even from Buffalo, Utica, or Rondout, N. Y., or from any other part of the world. Avoid all quack bitters usually sold at drugstores, as they contain alcohol. Avoid pills and pill-makers, for it is a well-known fact that the victims of pill-makers and pill-venders, whose bodies have been nearly destroyed by taking their vile nostrums, can be counted by thousands all over the world, wherever a maker or vender of these compounds has made his or her appearance. Possibly the inhabitants of the North Pole are not troubled with doctors, or pill-venders, and if so, that must be a paradise indeed, and when the hardy explorer opens the way, there he will be followed by millions of invalids who will be supremely happy to find a country where doctors and patent-medicine venders are unknown. I am happy to say, however, that the day is fast approaching when the healing art will be taught in the public schools of this vast Union, and then the army cf foolish nostrums, with the doctors, will only be known by reading the pages of a history of the past—excepting surgeons, who are the only members of this army useful to any community.

But, kind reader, excuse this digression. I promised to teach the art of healing and cleansing the blood, and curing all forms of disease, without the use of any medicine whatever, and I will proceed to fulfil that promise. I have one magic word which forms the basis of the whole—WATER. When its application is understood, it will cure every form of disease that the human family is subject to, and after an experience of thirteen years in its application, I have concluded to give

DIRECTIONS FOR TREATMENT.

In the first place, every person who desires to keep in good health should take what is usually called a hand-bath on retiring at night, at least once a week. This is done by dipping the hands lightly in hot water and rubbing the whole body briskly, then rubbing dry with a coarse towel. Remember this bath should always be taken on retiring at night, never in the morning. This treatment opens the pores of the skin, and cleanses the blood of impurities. Always rub the body with the hands, and not with a cloth or sponge. Very few persons realize how much they can do for themselves with their own hands, for every person has healing power. Very weak, or sickly persons, should be rubbed by some person or member of the family who is in good health.

Many cases of croup, lung complaints, throat or bronchial difficulties, can readily be cured by rubbing directly over the parts affected with the hands, dipping them in hot water, each night upon retiring. Lung complaints have often been cured by this treatment when all other remedies failed. A few spoonfuls of sweet cream sweetened with loaf sugar, and taken, is far better than cod-liver oil, as it imparts strength, and at the same time it is very cooling and healing.

Kidney complaints can readily be cured by this treatment. Heart disease also yields readily to this treatment. No remedy has ever been discovered, or ever will be, that will cure as many diseases as water will when used as I here direct.

For wounds and bruises, hot packing is the best, after rubbing the parts affected with the hands dipped in hot water. Wring out a cloth dipped in hot water, and put it on the wound, with a dry cloth over that. Avoid the use of any salve or plasters, as this treatment will cure quicker, safer and better than any remedy in the world. All forms of humors, running sores, heal under this treatment quicker than by any other yet discovered. Saltrheum yields like magic to this treatment. Every form of lameness and stiff joints yield best by this treatment. Swollen limbs and joints should not only be rubbed with the hands, dipped in hot water, but hot packing should be used also, as this alone will remove pain and inflammation. Cramps, severe pain in any part of the body, lameness in the side or back, can be cured by rubbing with the hands, dipped in hot water, and the application of cloths wrung out in hot water, covering with a dry cloth.

Bowel complaints yield readily to this same treatment. A treatment of the hands, dipped in hot water, will prevent a fever, cure a bad cold, and even break up a fever, although it be seated. Very fleshy persons, and persons inclined to dropsy, should resort to this mode of treatment, and avoid the use of potatoes, except sweet potatoes, as the white potato is very starchy and watery, which lays a sure foundation for dropsy. Every person should exclude pork from the diet, as pork alone causes more disease than any other article used; therefore its use should be abolished at once by all persons, young or old, who wish to enjoy good health. Persons troubled with any difficulty of the heart should at once discard the use of potatoes as food.

Any person who doubts the efficacy of water, let him test this treatment and be convinced of the truth of these assertions. If afflicted with catarrh, the following treatment will cure you: On retiring, inject a small quantity of water up each nostril, just enough to feel it in the throat; then take a thin cloth and fold two or three thicknesses, wring it out in hot water and place it directly over the eyes; then tie a dry handkerchief over that and keep it on during the night. This simple treatment will cure the worst forms of catarrh quicker and better than any known remedy, and the same treatment will cure the worst case of sore eyes. The sooner the use of medicine is abolished, and the clear sparkling water takes its place as a sure and safe substitute, the better it will be for the human family. You who are suffering from any form of disease or pain, apply this sovereign remedy and get sure and permanent relief, and thus test the truth of my statement, for water surely contains a divine life-giving principle truly surprising, especially to any person who makes its divine qualities a study and becomes a living witness of its magical effects. When properly applied, and its wonderful healing qualities are fully known and understood, and skepticism is felled to the ground by accumulated facts, posterity will erect a monument sacred to the memory of the student who has discovered the wonderful virtue and brought out the latent power in water to benefit mankind—a monument larger and more enduring even than the sacred monument of old Bunker Hill.

One more word in regard to the danger of
the loss of liberty in this county by and
through the doctors—not only doctors of medicine, but doctors of divinity also. The latter
would be very glad to have laws passed, compelling people to go and hear them preach,
while the former are equally desirous to have
laws compelling people to swallow their vile
compounds ; but the compounds of one are
as unwholesome and unpalatable the other,
although one is physical and the other spiritual. As the children of men advance in knowledge and truth they will refuse both. It would
be fully as just and proper for some railroad
company owning and running cars through a
state, to come before the Legislature of that
state and demand a special law forcing people
to ride on its road, claiming that all other
roads were not safe for the travelling public,
as for physicians to ask for special laws to force
people to employ them, making it a law punishable by fine and imprisonment for a common nurse to give an infant a dose of castor
oil, as she would then be practicing medicine,
according to the construction of these silly
laws passed by some of the United States.

One more word in regard to that magic word
the wonderful healing qualities of that truly
good beverage—WATER. Are you troubled
with costiveness, kind reader, as the result of
the baneful influence of medicine given by the
advice of some quack doctor? A tumblerful
of warm water taken before eating in the morn-

ing, every twenty-five minutes until a movement is effected, will cure you. Persons so troubled may usually take from two to five tumblerfuls before a movement of the bowels is effected. The worst case of costiveness can be effectually cured in a few days by this treatment. Are you deaf, or have you any difficulty of the ears—humors or running sores—rub the neck about the ears, briskly with your hands, dipped in hot water, on retiring. Take a small swab, dip it in warm water and rub the inside of the ear lightly, as far in as you can reach without hurting yourself. Rubbing the body two or three times a week, with warm water on retiring, will in a short time cure the worst form of liver complaint; and in fact any difficulty of the body, inward or outward, can readily be cured by the treatment herein described. Some persons say use cold water, but I wish to explain: cold water has precisely the same effect, with this exception, that the natural heat of the body must heat the water before any benefit can be derived, consequently if you have the water hot to commence with, it acts immediately.

In regard to cleansing the blood, if the body is rubbed once a week with hot water applied with the hands, on retiring at night, the blood will cleanse itself. But I have one word to say in regard to the false theory of quack doctors. To cleanse the blood by taking medicine is an impossibility, for there is no other process whereby the blood can be changed except by a proper regard to diet, and any impurity in the blood can only be carried out through the pores of the skin by keeping them open by rubbing the body, at least once a week, with hot water and the hands, on retiring.

If you are a user of tobacco, break off the filthy practice at once, for its continued use will surely shorten your life. To tone up the system to enable you to leave off, rub the body with hot water and the hands at least once a week, on retiring; also put a small handful of camomile flowers into a tumbler of cold water, let it stand a few hours, and of this take about three tablespoonfuls a day; this will tone up the system until the desire for tobacco passes away; and opening the pores of the skin by rubbing the body with the hands, dipped in hot water, will in a few weeks entirely eradicate this vile poison from the system.

Again, kind reader, don't allow any quack doctor to give you morphine, quinine, or opium, or anything containing alcohol in any form, to be applied outwardly or inwardly. Always remember that the physicians who use any of the above named articles in their practice must be classed with quacks.

Avoid all medicines of any description usually sold at drug-stores, for nature, used or applied as herein directed, will cure every ill that the human family is heir to.

Have you corns or bunions? Water, again, is your only relief. If on the toe, wind around it three thicknesses or more of thin white cotton cloth, fastened with woolen yarn. Each night on retiring, wet with water and let it dry through the night; wet again before dressing the feet in the morning. Serve a bunion in the same way. This bandage being thus kept wet with water, rots that hard unhealthy accumulation forming the corn or bunion, and promotes the growth of healthy flesh. In from ten to twenty days this treatment effects a perfect cure. This treatment also gives immediate relief from all pain.

For weak nerves, steep a handful of hops in water, rub the body with the decoction, also drink a few spoonfuls of it on retiring, and it will do you good. But do not use any of the vile preparations sold at the drug stores for such complaints, even though they may be recommended by Henry Ward Beecher, or any other learned divine. Preserve the health by the curative properties of water, and thus prolong your life. If troubled with dysentery, water is your remedy. Fill a large tub with water as hot as you can bear, get into it and remain about ten minutes, then rub dry with a towel. The extreme heat brings the inflammation outside and gives immediate relief. If troubled with piles, or any inflammation or trouble of any kind in the water or front passage —weakness or trouble of any name or nature— inject warm water once or twice a day. If there is much heat across you, use hot packing. Running sores, boils, and felons yield readily to hot water as herein directed, and will be found far better than any salve or ointment, especially those containing animal grease such as lard or tallow, which are rank poisons to all sores and humors. Follow carefully the instructions herein given and your health will be restored and preserved, money saved, and the services of the manipulator or magnetic doctor, who heals by the laying on of hands, can be dispensed with. There are plenty of persons about you who have healing power if they will only use it for the benefit of one another. I caution you against all persons who assume the title of mediums; if they use medicine, they are not true mediums. Carefully follow my instructions, and you will soon be convinced of the truthfulness of what I say.

Persons afflicted with chills and fever will find water to be their best remedy. On retiring at night put a great spoonful of red pepper in a quart of hot water, into which dip the hands and rub the body briskly, then rub dry with a coarse towel. Also drink a tea made by putting one teaspoonful of pure black pepper - -the berry ground or powdered—in one pint of milk, boiling it three minutes ; drink as hot as you can, at night and in the morning. This is a sure cure and also a preventive.

To prevent loss of the hair and to promote its growth, or cause it to grow where bald, dip the hands in hot water and rub the head thoroughly at least twice a day. This treatment promotes healthy growth of the hair where all

other remedies have failed, and it is the best
hair dressing in the world.

I would again caution my readers against
the use of pills of any name or nature ; also all
"Medical Discoveries,"sometimes called "Golden discoveries," for the gold, or the price you
pay for the vile compounds goes to enrich the
brazen-faced quacks who palm these preparations upon the unsuspecting public. Use
bright sparkling water as herein directed and
avoid injuring your health by taking any quack
medicines, or pain killers. Preserve the health
by following my directions, and be happy. If
you would be true to yourself, avoid clairvoy-
ant and medium doctors of any name and
nature, unless they are blessed with sense
enough to treat you without the use of medicine. Shun root and herb doctors ; your stomach was never intended for a dye factory, and
I assert that water contains more healing properties than all the roots and herbs that ever
grew. Keep yourself and your families in good
health by the use of water alone, and then you
will avoid being fleeced by doctors. If water
cost a dollar a bottle and persons would buy it,
turn a small quantity into a dish and rub their
bodies with it, they would obtain sure and
permanent relief far better than they do from
the vile compounds now in use. My friends,
do not take any stock in preparations put up
the name of any saint, for these old saints are
frauds, as old saints have been in every age,
and as history repeats itself, no doubt they

always will be. If you need a tonic, steep a few hops in water and drink of that ; or put a handful of camomile in tumbler of cold water and drink two or three spoonfuls a day of that and it will do you good; but avoid all preparations put up in alcohol under the head of bitters, and do not place any confidence in any of the remedies advertised in newspapers and recommended by some learned divine who knows as little about medical affairs as he does about spiritual. Use the bright, sparkling water, heaven's greatest gift to mankind, and you will secure happiness.

In writing a treatise on health, I find I have been so earnest that I have made a number of repetitions, almost to an earnest orthodox sermon; but I will let it go into print as it is, on the ground that truths as important as these cannot be too often repeated. As I have often said before, take good care of your body, as herein directed, avoid doctors and all medicines as you would shun a pestilence, save your health and money. During the past year a very learned doctor, who has spent over fifty years of his life in poisoning and injuring people's bodies with vile drugs and medicines, commonly used by the medical fraternity, and who has written several medical works which grace the libraries of many doctors, has very wisely come to the conclusion that if the practice of medicine had never been discovered, or come into general use, the human family would have been far better off, and that the average

length of life would have been far in advance
of what it now is. This is an honest admission,
but what a pity that this gentleman had not
discovered his great mistake earlier in life.
I wish here to say a few words in regard
to the extravagant fees charged by the average
doctor. It is a well-known fact that they become very rich after a few years' practice,
whilst a hard-working mechanic or laborer has
to work a lifetime to procure for himself and
family a home, often making a failure at that.
Hence the necessity for a general United
States law regulating the fees of the medical
fraternity, to bring their income down to a
level of the common laborer or mechanic who,
as I have said before, are the most useful in
society. Not long since, a hard-working mechanic near Norwich, Conn., had a child taken
sick with that dread disease scarlet fever. His
wife nursed the child and it soon recovered,
but having a second child taken with the same
disease, he became somewhat alarmed, and was
induced to summon a doctor. The result was
that the doctor managed, with vile drugs and
poisons which they usually use in such cases,
to kill the child in less than a week. The
mother firmly believes to this day that had
she nursed this child as she did the other, its
life would have been saved. For killing this
child, the doctor charged the hard-working
mechanic fifty dollars, a receipt for which the
poor man holds. I urge upon you, kind reader, to use your influence to procure the pass-

age of a general law regulating the fees of these vampires, that you and your neighbors may be protected from their unjust charges.

I again caution you against the use of alcohol in any form, except as heretofore specified. Alcohol given inwardly and used outwardly, in the case of our late lamented President, hastened his death, a fact that should be a warning to you never to use it or allow it to be used in your families.

If you wish to take a mild tonic, use camomile in water, or if you want celery, use it in its natural state ; and if you wish to use hops, or any of the plants or herbs, always use them in their natural state. Save your money and preserve your health by not allowing yourself to be imposed upon by medicine venders, who puff the qualities of their vile goods under the names perhaps of camomile or celery pills, or bitters, or ginger tonic, and other vile preparations prepared with alcohol. Follow my directions and you will surely enjoy good health and be happy. Do not forget, as I have often told you before in this work, that the oftener you see any medicine of any name or nature puffed in the newspapers, the greater the humbug, and the oftener you see the homely profiles of medicine venders—whether of men or women—the greater the humbug. Bear this in mind, kind reader, for the truth will always bear repeating.

One little incident more showing the necessity of a general United States law regulating the fees of doctors, to prevent them from robbing people as they are now doing throughout the entire length and breadth of our land—robbing alike both rich and poor—and I will bring this work to a close, and tax your patience no further. Other countries have been obliged to do this to protect their subjects, then why should the American Government do less for theirs, for it is the sacred duty of all governments to protect their subjects.

Not long since, a gentleman of my acquaintance, somewhat advanced in years, was taken ill, and it was apparent that this would be his last sickness. It is very fashionable to employ some learned ignoramus of a doctor to aid a person to die, and this old gentleman followed the fashion. The result was that the doctor helped him off in about six months' time. If the doctor had been kept away from him, it is quite evident that he would have lived at least a year. At the time he was taken sick he was worth eleven hundred dollars, one thousand of which was paid for medical services, leaving but a mite for the poor worthy widow, who was obliged to spend the remainder of her days in the poor-house. This is only one case out of a thousand that transpire in our midst nearly every day of the year—doctors robbing widows and orphans of that which is justly theirs—plainly showing the immediate necessity of a general United States law to effectually remedy and suppress this great and growing evil.

CONTINUATION.

After looking over this work, I have come to the conclusion that perhaps I have been too severe upon the medical fraternity, as I have repeatedly spoken of quacks; consequently, any practitioner of medicine need not put on the coat unless it fits him. All doctors using the vile compound called lager beer or any preparation containing alcohol, or quinine, morphine, opium, calomile, must be classed with the lowest order of quack. Although, as said before, the above named articles my give temporary relief, they are injurious and dangerous to the patient, leaving in the body a permanent injury; consequently the truly honest physician will not use them in his practice; but the less medicine he uses and the more he studies and understands the curative of bright, sparkling water the better will be his success, and the greater benefactor he will be to mankind.

Are you, dear friend, troubled with kidney complaint? Take 1 pint of blueberries, or common black whortleberries, steep them one half hour; strain through a thin cloth seive, sweeten with loaf sugar, and take three tumblerfuls each day. Each night on retiring rub your body with hot water, rubbing briskly with the hands. A few days of this treatment will cure the worst cases of kidney complaint.

If you are troubled with piles, by taking pills, or from any cause, each night on retir-

ing take an injection of warm water; have a
large tub or vessel filled with water as warm
as you can bear it, and sit in this for ten minutes. A few days' treatment as thus directed,
will cure the very worst cases of piles. If you
are subject to headache, on retiring at night
rub your stomach with your hands, dipped in
hot water; also rub the feet, and the legs from
the knees down, taking special pains to rub the
bottom of the feet. This treatment will cure
the most obstinate or severe headache. For
chronic diarrhœa. have a bandage made of
cotton flannel or some thick white cloth; have
this bandage wide enough to cover the bowels, and made so as to go around you and
button; once a day rub yourself around the
waist with your hands, dipped in hot water;
take a thin strip of white cotton cloth, one
inch narrower than the thick bandage, dip it
in hot water and wring out, place it over the
bowels and place over it the thick bandage,
doing this on retiring at night, and again in
the morning. For diet use gruel made of oat
meal, or flour, or both, if you prefer. Use
good new milk in your gruel. Eat toast or
crackers if you prefer. Do not eat any pork.
This treatment has cured many cases of this
malady where all other remedies have failed.
Use all the good sweet cream you like, sweetened with loaf sugar.

I notice that some learned fool recommends
invalids and rheumatic people to wear red flannel underclothing. I wish to impress upon

the mind of every reader of this work that that
is one of the greatest mistakes of this enlightened age. No doubt this statement will surprise some invalid who has become red-flan-
nel crazy, but I will explain the reason why all
persons, sick or well, should avoid red flannel.
Those who wish to enjoy good health should
wear white underclothing always, for white is
a conductor of electricity, while all or nearly
all colored goods are non-conductors; besides,
in all red flannel, and nearly all colored goods,
deadly poisons are used in making the dyes,
as many persons can testify whose bodies have
been poisoned by wearing such clothing.
Therefore, always wear white clothing next
to the skin. Not many years ago a set of
medical swindlers bought a million or more
of white under-shirts and drawers, at less than
twenty cents per garment, run them through a
vile red dye, and then sold them at two dollars
or more each, claiming they were medicated.
This was the "boss" of all frauds. Do not take
any stock in medicated garments sold at very
high prices, as lung protectors, liver pads, etc.,
but protect your lungs and liver with nice
white flannel.

If you are chilly, or have a cold, and wish
ginger tea, buy some Jamaica ginger root, and
it will do you good, but avoid as you would a
pestilence the vile preparations called Jamaica
ginger. For a tonic use camomile; for a nerv-
ine to strengthen your weak nerves, use mild
hop tea. Do not forget the golden number

three: eight hours for work, eight hours for
rest and improvement of the mind, and eight
hours for sleep—three times eight are twenty-
four. This division of the twenty-four hours
must eventually become the universal law of
this entire Union. Such a division would
prove one of the greatest blessings to mankind.

LIVERYMEN, LACKEYS AND COACHMEN: OR THE SKEL-
ETONS OF DEPARTING AND DEPARTED GREATNESS.

Within the last few years, many philan-
thropic societies have been formed to prevent
cruelty to men, women, and children, as well
as to animals. While teaching people to heal
themselves without the use of medicine, I have
passed a number of seasons in Newport, R. I.,
and Saratoga, N. Y. I have frequently noticed
the extreme cruelty of the rich aristocrats—
these relicts of departed greatness—requiring
their women servants to work in low, damp
basements, giving them little time for recrea-
tion and rest. Such treatment soon ruins the
health, and the victims frequently pay all their
earnings to some quack-doctor, and every dose
of his medicine that they swallow only serves
to make their case more critical. But these
societies will soon teach these aristocrats that
they must build their kitchens above ground.
Plenty of time to bury the poor servants after
they are dead, without burying them alive in
damp, unhealthy *kitchens*. In regard to liverymen and coachmen, they are compelled to
shave their faces nearly every day, thereby

causing injury to their health by not allowing the beard to grow on the chin and neck for a natural protection. Such cruelty should not be tolerated, and these rich tyrants will ere long learn that they do not own the gentle-men, their servants.

Surgeons for horses will always be a neces-sity; but the numerous quacks advertising to be horse-doctors, will ere long attract the ear-nest attention of the Society for the Prevention of Cruelty to Animals, when these quacks will no longer be allowed to punish the poor horses. Even though their remedies may afford tem-porary relief, they eventually destroy the con-stitution. Give the horse good care and good food; and if he gets sick, give him rest, but never allow him to be dosed by quacks. The less medicine given a horse the longer he will live, and the better service he will render to his owner.

————o————

ANSWERS TO EVER-OCCURRING QUESTIONS.

After writing the foregoing, and stating the facts and giving directions so plainly that I thought no reader could possibly misunder-stand, I have been asked many questions which I will here give, and answer them more min-utely, if possible, than I have in the preceding pages of this work.

Some readers tell me that while I do not dis-card medicine wholly, I have about the same as said that it has never been of any earthly use, thus disputing the so-called wisdom of the whole army of medical practitioners of the past. I again assert that pure sparkling water, properly applied, will cure every disease that man or beast is subject to; and that all roots, herbs,vegetables,plants,vines, minerals,drugs, gums, often prepared in plasters, and all preparations that are now, or ever have been, used by the medical faculty, are not of any earthly use, compared with water; whilst water, when properly applied, makes a perfect cure of every curable disease,leaving the patient in a healthy and hardy condition. All the rest, whilst they do sometimes give temporary relief and apparently help in one direction, are sure to injure in another. Let all medicine venders, medicine takers, and the medical faculty with all their so-called wisdom, dispute this plain, simple fact, if they can. One ancient writer has very truly written that man was made upright and good, and although he has sought out many inventions, many of which have been a blessing to himself and the world, the most injurious, worthless, and useless, has been the invention of medicine, when compared with the divine curative properties of water, applied as I herein direct.

One person tells me that I have not given directions quite plain enough in regard to that throat difficulty called diphtheria, sometimes called quinsy. For this take one great spoonful of sharp hot water, as hot as you can bear without burning, every fifteen minutes. For a small child, one teaspoonful every fifteen minutes. Rub the neck and throat faithfully with the hands, dipped in hot water. Take a light, thin cloth, folded two or three thicknesses, wring out of hot water, and put it around the throat with a heavier cloth over it. Repeat as often as the inside cloth gets dry. This treatment is the best and surest to give relief of any that has ever been discovered. Also rub the feet twice a day with the hands, dipped in hot water.

Another person asks if the plasters sold at drug-stores, usually punched full of holes, are good. I answer that they are of no earthly use. If they relieve in one way, they will be sure to injure in another. In place of plasters, rub the affected part with the hand dipped in hot water. Wring out a thin cloth dipped in hot water, and put on instead of the plaster, covering the wet cloth with a heavier dry one. Wet the inner cloth often as it gets dry.

Another thinks the dyspepsia cure not quite plain enough. Each night on retiring rub the stomach and back briskly with the hands dipped lightly in hot water; also rub the feet in the same manner. The last thing on retiring, drink about one-half teacupful of sharp warm water, and another before eating at each meal.

For cramps, or severe and sudden attacks of pain, rub briskly the parts affected with the hands, dipped in hot water. Lay on a light cloth, wrung out in hot water. If this does not relieve, heat an earthen plate in the oven or on the stove, and place it over the part affected. Any bruise or sore having what is termed proud flesh in it, should be treated with the hot plate for a short time, as the extreme heat thus obtained acts as caustic. Any sore on the body, where caustic is usually applied, can be treated as I here direct with success, and is not attended with as severe pain, besides leaving the patient in a much better condition, and effecting a more permanent cure. Persons subject to fits or spasms should take a hot bath, getting into a tub of water hot as possible—doing this at least three nights in a week on retiring. Remain in the bath about ten minutes, then rub briskly with the hands for a few minutes, and rub dry with a coarse towel. This treatment is a cure, and also a preventive.

Still another person says, that 1 am injuring his business by giving this knowledge of healing broadcast, as it were, to the world. To which I simply reply that the motto is: "The most good to the greatest numbers." Another says he is a healing medium, and gets his living healing sick people by laying on of hands, usually called magnetic treatment. In reply I would say, that giving these treatments constantly is very weakening, shortens the life, and takes away the vitality of the person giving them, and does not do justice to the person receiving the treatment. By giving many treatments,the person receiving them becomes weaker and more exhausted than they were before being treated; consequently, this work should be divided up by each one helping themselves and helping others, and not throw all this burden upon a few.

Still others ask me what about faith prayer cures; seventh sons and daughters. To these I say that the power of mind over matter is a wonderful power and not fully understood at the present day. If two or three persons assembled around a diseased person wish to give or transmit their strength or vitality to the sick one by prayer, I have no objection; yet it can be accomplished equally as well by these persons gathering around the patient, being equally as earnest for the restoration of the patient to health and strength. It is wholly a power of the mind. In regard to seventh sons and daughters having any healing power more than any of the other sons or daughters of any family, is an old superstition theory of the past, for any one of the others has the same power and ability to heal as the seventh, provided they are equally as strong and healthy, and try as earnestly so to do, with the same amount of faith on the part of those desirous of being healed.

Another wants to know how I dare speak so decidedly against the combined wisdom of theologians who believe in and practice using a little whiskey and wine for their health; and again, how I dare speak against the one old saint of whom we read in some rather doubtful history, who advised another old saint, who doubtless had the dyspepsia by being overfed, as many saints are at the present day by the lambs of their flock. It seems that one advised the other to take a little wine for his stomach's sake. Whether this other old saint took the advice or not history has failed to inform us; but I infer that he did not take such foolish advice; my advice would have been to take a little hot water for his stomach's sake, and not eat quite as much. Science sustains me in saying that all alcoholic drinks are dangerous and hurtful when used either as a beverage or as medicine, internal or external. Another wants to know why I discriminate against pork and not other animal food. Pork produces so much heat in digesting, that it not only makes persons naturally full of humor much worse, but also has a tendency to produce humors. What food is the best asks another. Eat that which agrees with you best and reject that which you do not want or that injures. If a public speaker or singer, and wish to strengthen your voice, rub your throat each night on retiring, with your hands dipped lightly in warm water, and as you lie down drink a swallow of warm water. This simple treatment will strengthen the voice better and give more permanent benefit than any medical preparation ever made or that ever will be.

Are you troubled with constipation caused perhaps by vile compounds taken by advice of some fool of a doctor? Each day before eating drink one teacupful of sharp, hot water, and each morning have a nice good orange cut up fine and sprinkled over with sugar, and eat it after drinking the water, before eating anything else. Eat one at each meal if one at breakfast does not have the desired effect. If you have a bad cold, break it up by the use of water as herein directed, as this method will give sure and permanent relief.

Are you troubled with dropsy, brought on perhaps by eating potatoes, or by hard work, or by taking driving or scattering medicines, which have caused your complaints to assume a dropsical form? Water is the only sure cure. Drink one teacupful of water hot as you can bear it, before each meal; rub the body at least three times a week, on retiring, with the hand dipped in hot water; chafe dry with a coarse towel; each night on retiring rub the parts of the body swollen most with the hands dipped lightly in hot water. If there is too much heat, use hot compress or packing. Hot water is an absorbent, and when used as here directed will cure the worst forms of dropsy better than any remedy ever yet found or that ever will be. Medicine takers, note this fact.

For measles, take a few swallows of hot water every half hour; once a day take a hot-water bath, remaining in the water ten minutes, with heat up to from 85 to 102 degrees; chafe dry with a coarse towel. Continue this treatment until the measles are fairly out, and continue to drink hot water until recovered fully. Use this same treatment for small-pox, and to break up all kinds of fevers; but for yellow fever drink the hot water every ten minutes, and take the hot bath every six hours. A sure cure for cholera: The hot bath should be taken every two hours, and the patient should drink all the hot water possible.

For consumption or lung complaints, one-half hour before eating drink one pint of sharp hot water; eat beefsteak done rare, with a very small slice of stale bread. Nothing else. For rheumatism, let some strong person rub you, on retiring, rubbing the parts affected with the hands dipped in hot water; drink a tea-cupful of hot water before each meal. For dyspepsia drink a teacupful of hot water before each meal, and on retiring rub the stomach with the hands dipped in hot water. Rub briskly for cramps in the stomach or a stoppage, and drink a cup of hot water every ten minutes until relieved. For neuralgia, rub the parts affected briskly with the hands dipped in hot water ; if this does not give the desired relief, use hot packing. Sometimes hot water run from a sprinkler has a good effect. A cup of hot water should be drank every half hour.

No person taking a hand bath once a week, as herein directed, can ever become bilious, or have a fever. Take this weekly bath, and drink no cider, wine, beer, or anything containing alcohol, it is a sure preventive against small-pox or yellow fever. Never allow yourself or any of your family to be vaccinated to prevent small-pox, as vaccination is highly dangerous, and is only another plan of medical swindlers to tax and defraud the people ; instead of being a preventive, it lays a grand foundation for many other diseases. Instead of being a great public plessing, as the medical faculty would have you believe, vaccination is a stupendous curse. Instead of being made compulsory by law, as at the present day, the time is not far distant when every intelligent community will prevent it by law, and when the use of pure, sparkling water is fully understood, the medical societies, medical swindlers, and medicine takers will only be known by reading the pages of history. We shall then need no laws to regulate the practice of medicine, and shall hear no more quarrels about the issuing of diplomas. It is difficult to tell which are the greater set of frauds, the regulars, or what are termed the irregulars, they are all such contemptible humbugs it is almost impossible to discern the difference between the two.

For liver or heart difficulties: on retiring rub the body briskly with the hands dipped lightly in water as hot as you can use it chafe dry with a towel, and drink a cup of hot water before each meal. By rubbing any diseased part of the body with the hands dipped in hot water you get the healing power of the hands and hot water combined,

making two of the best healing agents in the
world.

Persons troubled with diabetes have found
water to give the best relief. Drink a cup of
sharp, hot water before each meal, and rub
the body with the hands dipped lightly in
hot water each night on retiring. Eight or
ten days of this treatment usually effects a
perfect cure.

A treatment of hot water has always
proved to be the best remedy in the world
for female difficulties and weakness of any
name or nature, connected with the water
passage. For any of these difficulties inject
water as hot as you can possibly bear it two
or three times a day; at the same time rub
the parts affected briskly with the hands
dipped in hot water. Also take a thin cloth,
wring out of hot water and place over the
affected parts. Renew the hot compress or
packing as often as the heat of the body dries
it. This treatment is the safest, quickest, and
best remedy ever found for any of these diseases.

For general weakness or irregularities
caused perhaps by taking cold or from any
any other cause, drink a pint of sharp hot
water on retiring, and take a sitz bath, sit-
ting in a tub of hot water, hot as you can
bear it without burning for about fifteen or
twenty minutes. This must be done at the
end of the month, or at the time for the flow
of the menses. Ladies desirous of bearing

children should not take a sitz bath for the first two or three months after being in that condition, for a sitz bath at the end of either the first or second month would in nine cases out of ten be almost sure to change that condition, which in time would be very weakening to the body.

All difficulties of the back or front passage caused by piles, often brought on by taking pills or bitters, which in nine cases out of ten produces piles, can readily be cured by taking a hot water injection and sitz bath at night on retiring. Any discharge or irritation outside or inside of front or water passage can be cured in a few days by using injections of water as hot as possible without burning. Each night on retiring take a sitz bath. Eight or ten days of this treatment usually effects a perfect cure of any of these troubles. Avoid all strong drinks and do not eat salt or greasy victuals. Use the hot compress, too, during the day, and drink plenty of warm milk.

For pimples on the face, rub the body at least twice a week with the hands dipped in hot water, on retiring, and always wash the face in water as hot as you can bear it. Never use soap in the water for any disease, and avoid all extracts of any name or nature. Spring water, or nice clear water from a pond, is the best extract in the world.

THE MANY ADVANTAGES GAINED
BY THE USE OF HOT WATER.
———o———

Softening of the brain has often been cured by taking a full bath on retiring, with water as hot as can be used without burning, remaining in the bath ten or fifteen minutes, and then rubbing the body dry with a coarse towel. Drink a teacupful of hot water before each meal, and take this bath every third night on retiring. For asthma, drink a cup of hot water before each meal, and take a hand-bath every third night on retiring; rub the throat and breast every night with the hands dipped lightly in hot water, rubbing it in as you would a liniment; wring a light cloth out of hot water, and place directly over the throat and lungs, letting the heat of the body dry it through the night; rub the feet and legs from the knees down, taking special pains to rub the bottom of the feet.

Paralytic persons should be treated in the following manner: Let some strong person give them a hand-bath each night before retiring, dipping the hands lightly in hot water, rubbing it in as you would a liniment; rub the body dry with a coarse towel; drink a cup of hot water before each meal to cleanse the blood, or as a tonic for the stomach. Drink a teacupful of hot water before eating in the morning, to break up a cold.

Drink a pint of hot water on retiring at night, to break up a bad cough. Drink a cup of hot water two or three times a day—is one of the best remedies in the world for any stomach trouble, and to remove phlegm or any unhealthy matter from the stomach or intestines. For whooping-cough, take a great spoonful or two of hot water every half hour during the day, and drink a few swallows whenever there is a hard coughing spell. For mumps, drink three cups of hot water each day; wring a light cloth out of hot water, and place around the neck and face, covering with a dry cloth, and renew as often as the inner cloth gets dry. If troubled with severe pains in any part of your body, rub the parts afflicted with your hands, dipped lightly in hot water, and drink a cup of hot water; wring out a cloth in hot water, and place two or three thicknesses directly over the parts afflicted; heat a soapstone or brick, and lay outside against it, as hot as possible, without burning —this will give immediate relief. For tooth-ache, hold hot water in the mouth, and rub the gums and the face with the hands, dipped in hot water; if this does not give the desired relief, wring a cloth out of hot water, and place over the afflicted part; also heat a soapstone or flatiron, and lay the face against it—as hot as can be borne without burning; this gives relief when all other remedies fail.

Always use hot water for weak or sore eyes. Never take any medicine from any doctor for this trouble, for the doctors who treat for diseases of the eyes give the most deadly poisons. To banish worms from children or grown people, drink freely two or three times a day of hot water, and three nights in the week take a hand-bath, dipping the hands lightly in hot water, rubbing the entire body briskly, then rubbing dry with a coarse towel. This treatment creates a healthy condition of the stomach and intestines, and banishes every species of worms. For the bite of a mad dog, or to prevent hydrophobia, wash the wound immediately with hot water; give the patient all the hot water he can drink the first day; give a full bath, as hot as possible, remaining in the bath one-half hour; use a hot compress on the wound until healed; take the hot bath every day for a week; drink three cups of hot water a day for twenty days. Treat the bite of poison snakes the same. In any part of the body where it is necessary to use the hot compress to retain the heat, thin sheets of cotton batting can be laid over, covering with oil silk. Avoid medicine, and never take any preparation of wine or iron, for it is a very dangerous preparation, and very injurious to the body. You may rest assured that you get all the minerals you need in the food you consume. Don't let any medical quack fool you by telling you that your body needs iron, and do not employ any doctor who has not common sense enough to treat you without medicine; and always remember that medicine doctors and State medical associations of doctors, are nothing more nor less than medical frauds; and always bear in mind that this so-called medical science is the most unreliable science in the world, and only third or fourth-rate doctors ever use medicine in their practice.

For croup, give the child a teaspoonful of hot water every four minutes; rub the throat and lungs with the hands, dipped lightly in hot water; use a hot cloth across the throat and lungs, and keep the feet warm. For rash, or breaking out on any part of the body, take a full bath, with water as hot as possible, without burning. Treat chicken-pox the same as measles, and always use hot water for all sickness of horses and cattle. Obey these instructions, and you will be happy by using the only universal panacea ever discovered for curing all diseases that man or beast is subject to. Very many persons prefer to use cold water for baths, but if the water is used cold, the body must heat it before it can have any action whatever; therefore always use it hot. But in case a person was placed where hot water could not be had, a good result might be obtained by dipping the hands lightly in cold water and rubbing the body briskly, so as to produce heat immediately on the body or parts afflicted. For heartburn, drink a cup of hot water, and rub the sides and breast with the hands, dipped in hot water. Treat heart-troubles in the same manner, and avoid eating potatoes, except sweet ones—they are a great cause of dropsy and heart-disease.

——o——

CARE OF INFANTS OR SMALL CHILDREN.

No medicine should ever be given to infants, for medicine, if it helps in one direction, will be sure to injure in another. All medicines are equally injurious to children as well as to grown persons. Avoid as you would a pestilence all medicine, either for yourself or your children, whether they are called " Safe cures," or by any other name. The safest and surest cure in the word is water, when used as I have herein directed.

While children are teething the face and gums should be rubbed every two or three hours with the fingers dipped in water as hot as can be borne without burning. No child should be permitted to nurse more than three times in twenty-four hours, and if troubled with cramps or pains in the stomach or bowels, one dessert spoonful of water as hot as can be used without burning should be given once in four minutes until relieved. In the case of small infants, the whole body should be rubbed three times a week with the hands dipped lightly in warm water. If a child is restless give a few teaspoonfuls of water as hot as possible without burning, but in the name of heaven never allow to be given to your children any of the miserable preparations called "Soothing Syrups," said to be put up by some old Christian mother, for they are universally stupendous frauds, and have ruined and killed more little innocent infants than any other remedy ever used or heard of. If any person could take a look through the country for the dear old mothers who are said to put up those vile preparations, they would find some old bloated red-faced Dutchman who would hardly be able to distinguish an infant from a haystack.

Avoid all these things against which I have warned you in this book, and you will preserve your health and save your money. Be happy in using the only true cure ever yet discovered by mortal man for curing all diseases of the children of men.

DIRECTIONS FOR TREATMENT OF INSANITY, FEVERS, RUNNING SORES, AND CANCERS.

For any form of insanity, give one teacup of hot water to drink three times a day. Two nights during the week, or every third night, on retiring, give a full bath in a tub of water, hot as possible without scalding. Let some strong, healthy person rub the whole body with the hands dipped lightly in the hot water, rubbing dry with a towel.

Keep the patient in the hot water from ten to fifteen minutes for scarlet or typhoid fever. Give one-half teacupful of hot water to drink

every hour. Rub the whole body once a day
with the hands dipped lightly in hot water.
Wring thin cloths out of hot water, and lay
across the bowels and groins. Renew the
hot compress as often as the heat of the body
dries it. This treatment has often broken
the worst forms of this fever in three to four
days.

For lung-fevers, place the hot cloths over
the lungs and across the back and shoulder-
blades. Rub the whole body once a day with
the hands dipped in hot water. Administer
one-half teacupful of hot water every half
hour. Oilcloth can be used over the hot
compress, if the heat of the body does not
dry it or the person is chilly; or a hot brick
or soapstone can be used if there are chills.
This treatment usually breaks up the worst
forms of lung-fevers in three to four days.

For spinal meningitis, administer one-half
teacupful of hot water every half hour. Rub
the whole body once a day with the hands
dipped lightly in hot water, taking especial
pains to rub the back and across the shoulder-
blades; also use the hot compress the entire
length of the back, and especially across the
shoulder-blades. Keep the feet warm with
hot soapstone.

For croup or asthma, or any difficulty of
breathing, give one teacupful of hot water
three times a day. Rub the whole body with
the hands dipped in hot water. On retiring,

place a hot soapstone at the feet. Take especial pains to keep the feet warm.

For cleansing the blood, drink one teacup full of hot water before eating, three times a day. Twice a week, on retiring, rub the entire body with the hands dipped lightly in hot water ; chafe dry with coarse towels. This is the best blood purifier in the world.

For cancer drink three cupfuls of hot water a day, one before each meal; rub the whole body twice a week, on retiring, with the hands dipped lightly in hot water ; dress the sore with thin cloths, covering with oilcloth to retain the heat. Renew the hot compress as often as the heat of the body drys it. If there is proud flesh or impurity, throw hot water for ten or fifteen minutes into the sore with a syringe once a day, or it can be run on with a sponge, but it should always be used as hot as possible without scalding. Dress all humors or running sores on any part of the body in the same manner. This is the best dressing in the world and has often healed cancer and running sores of long standing where all other remedies have failed.

For rash or any breaking out on any part of the body, or any skin disease if all over the body, take a full bath each night on retiring until it disappears. Two or three days of this treatment usually effects a perfect cure. Drink a cup of hot water before each meal. Take the bath as hot as possible without scalding.

For dysentery or any form of bowel complaint, or bowel complaint sometimes called bloody flux, drink a cup of hot water every half hour. Get into water as hot as bearable, remaining about twenty minutes. This treatment gives immediate relief. Take a full bath every three hours until relieved, which will usually be from twelve to forty-eight hours.

For scrofula or syphilitic sores on any part of the body drink hot water, one teacupful before each meal; take hand bath every third night on retiring; treat the sores same as cancer sores. This treatment cures the worst forms of these diseases, with a few weeks' treatment, better than any remedy ever yet discovered by mortal man.

Obey these instructions laid down in this book, my friend, and don't put confidence in any doctor or nurse, no matter from what part of the world they hail, or how great a reputation they may have. If they use medicine of any kind in their practice, they will be of no account whatever. Don't put any confidence in root and herb doctors. All the roots and herbs we need are those used for food.

ADDITIONAL INSTRUCTIONS.

Persons afflicted with St. Vitus's dance and other nervous diseases should take a full bath in water hot as possible, without burning, on retiring at night, three times a week, at least, and drink a teacupful of hot water before each meal through the day. Eat good, nutritious food, and avoid taking any nervines or medicines of any name or nature.

For crick in the back or any lameness in the sides, or any form of kidney, or any trouble on the sciatic nerves, rub the parts afflicted on retiring at night with the hands dipped lightly in hot water. If this does not give the desired relief, wring cloths out of hot water, put on, covering with dry ones; use a hot soapstone or hot rubber bag if the body lacks heat or you are chilly.

For chills and malaria drink a cup of hot water every fifteen minutes. In sections where chills and fevers prevail drink a cup of hot water as hot as possible without burning three times a day before eating. This is a sure cure and also a preventative. Take a hand bath every third night on retiring. Use the water hot as possible without burning.

For kidney trouble, lame back, or lame sides, great benefit can be derived by having a band of light cloth about twenty or twenty-four inches wide; wring out of hot water, place around the entire body. Wear it night and day, wringing it out of hot water as often as the heat of the body dries it. You can fasten it around the body with pins. This alleviates pain and is much safer and far better than any plaster ever made. Any person having caught a sudden cold or rheumatic pains whereby the entire body is in great misery, will find great relief by drinking a teacupful of hot water every twenty minutes, setting the feet in hot water above the ankles and holding the hands in a dish of hot water above the wrists. This treatment gives immediate relief.

Persons in that destructive habit of taking morphine, opium, or any of these vile nervines so destructive to human life will find hot water their only relief. Take a full bath each night on retiring, and drink a teacupful of hot water every hour or oftener if in great pain. A few days of this treatment faithfully adhered to enables you to break up this vile habit.

Tobacco users and persons addicted to using strong drinks have found this treatment to be the only safe road out of any of these destructive habits, and when that extreme desire returns for strong drink, hot water drank freely has been found the only sure agent that would give relief and take away that appetite for alcohol.

In regard to purification of the blood remember that there are only two agents in the world that can, under any circumstances, purify the blood—one is the breathing of pure

air and the other water; but when taken
cold the body must heat it before it acts,
consequently if taken hot it acts immediately,
for heat is life. To many persons resting
upon the false theory of medicine vendors in
regard to the curative qualities of safe cures
made up of iodide of potassium and gum guiac,
sarsaparilla, with other vile preparations put
up in alcohol, I wish to say distinctly that
none of the roots or herbs have, or ever did
have, any of the healing qualities claimed for
them by the average medicine vendors under
the vile name of Indian preparations, for history very plainly informs us that the Indian,
when first discovered by white men by Co-
lumbus and others, did not use any roots,
barks, or herbs as medicine, and not until
some of their numbers had been carried to
Europe and studied botany, did they know
anything about using any roots, herbs, or
barks for medicine in any form. They used
them only for food which contained nourish-
ment. Shun all medical preparations, and
bear in mind that all the vegetables we need
or have any earthly use for are those used as
food, and the more nutrition any vegetable
contains the more valuable it will be as a
medicine.

Again, my friend, are you troubled with
bleeding at the lungs or stomach, or any
gatherings inside, ulcers, tumors, or anything
like heartburn or any irregularities inside?
Hot water is your only sure, safe relief. For

any raising of blood drink a teacupful of hot water every hour, or if very weak, take a tablespoonful every few minutes, and lay a thin cloth wrung out of hot water directly over the parts where the blood seems to proceed from, covering with dry cloths to retain the heat.

To equalize the circulation of the blood if chilly, always use a hot soapstone or hot water rubber bag to retain the heat.

For brain fever drink all the hot water possible, or a few swallows every few minutes, and keep hot cloths on the head. Use the soapstone or hot rubber bag both for the head and feet.

For a boil or felon always dress with hot cloths, holding the hand in hot water. To run the heat beyond the natural heat of a felon will usually prevent it from coming. Always dress all gatherings of this nature with hot water and hot cloths, as this treatment will alleviate the pain, cause them to gather and break better and quicker, and heal them up in better shape than any preparation in the known world.

Again, my friend, don't pay any attention to those vile preparations under the name of tonics and sarsaparilla, for this root has never had any such qualities for purifying the blood or strengthening the body as claimed for it by vile medicine vendors, for nutriment alone can strengthen your body, and hot water is the best and only universal health renovator in the world, and if used as herein directed will effectually check all epidemics and prevent catching contagious diseases ; and the time must eventually come when this greatest of all curative agents will be used in all public hospitals, and all drugs, minerals, poisons, roots, herbs, and other vile preparations now used or that ever have been used by the medical faculty must be forever dispensed with, and honest taxpayers will no longer be taxed to run these public hospitals and other institutions now run in the interests of vile medicine vendors and dispensers of vile drugs, and all honest people will refuse to pay their money for any newspapers that are used to puff the vile qualities of medicines and medical preparations which never existed only as statements made up of printers' ink and wholly untrue at that. Let all invalids and weakly persons obey nature's laws as near as you can. Use this greatest and best of all curative agents, and your health will be restored and you will enjoy happiness.

Are you troubled with numbness of the arms or legs or lack of circulation of the blood which causes sleepiness? Drink a teacupful of hot water before each meal. Rub the arms and shoulders each night on retiring with the hands dipped lightly in hot water. Some people think they get a better result by drinking the hot water one half hour before eating, but every person must be

a law unto himself. Use it the way that it
does you the most good. If you are very
fleshy and would like to reduce your weight
drink one pint of hot water one half hour before eating. Eat stale bread and dried beef
dipped in cold water. Eat beefsteak and
other lean meats, but do not eat fat meat.
This diet will enable you to retain your
strength while the flesh wears away. This is
the safest, best method ever yet found.
Under this treatment the flesh usually wears
away at the rate of twelve to twenty pounds
a month. For severe cramps in any place on
the body get into hot water all over. Also
drink a cup of hot water every few minutes
until relieved, or use hot cloths wrung out of
hot water with soapstone or hot water bag
outside.

HINTS TO INVALIDS AND NURSES.

Always remember and bear in mind that medicine and medical doctors are absurdities. Although persons may study medicine according to the custom of the times in all sincerity, when they arrive at years of understanding they will soon learn a better way to cure disease than using vile medicines of any kind; and any man or woman, after having arrived at that age in life, and still continuing to dose the sick ones with medicine, must always be classed with knaves or fools. If they themselves have faith in the curative qualities of medicine, which never have existed and never will exist, they must be classed with the fool; but if with all the enlightenment of this age, they persist in dosing the poor victims who chance to fall into their hands with medicine, they must be classed with knaves. And any person who supposes that the doctor who uses medicine ever has cured, or ever will cure, anybody or any disease, makes the greatest mistake of his life. A person by the action of nature may sometimes recover and give the credit to the medicine, where it does not properly belong, for it is a well-known fact that in all past ages associations of medical doctors and medicine-venders have always been sad failures, and doubtless ever will be, and whenever a book, bill, or paper comes into your houses, or into your residences, puffing the qualities of any medicine of any name or nature, use them at once to kindle your fires with ; reduce them to ashes at once, or put them where they will do the least possible harm. Teach your children and all your household to shun all medicines, no matter from what source they may come, or who recommends them ; whether they are recommended by persons in private or public life, retired doctors of divinity or any other doctors ; but always learn to cure yourselves, and all whom you meet in every walk in life, by the divine, all-curative qualities of water. But if used cold, the body must be able to heat it before it will act; therefore study the best and safest method of applying it, heating it before using. For any irregularity of the system or gastric trouble, rub the entire body on retiring at night with the hands, dipped lightly in hot water ; drink a teacupful before each meal, and one more on retiring at night, and any time you chance to eat a late supper, drink a cup of hot water before retiring, and you will be all right in the morning, as this will prevent a late supper from doing you any harm. Persons addicted to drinking alcoholic drinks should live on a vegetable diet, and avoid meats awhile, and drink a cup of hot water whenever that desire for strong drink annoys them. Take a hand bath at least twice a week on retiring at night. To take the poison in the body out, or eradicate it from your system, whether it has come there by the use of tobacco or alcohol, or by taking poison medicines administered by some enlightened medical doctor, drink about one pint of hot water on retiring, take a cotton sheet, wring it out of hot water, and wrap it around your entire body, next to the skin. Wind woolen blankets or comforters around your body, and let the heat of the body dry it. Drink a cup of hot water before each meal. Three or four nights of this treatment on retiring will eradicate the worst of poisons from the body. A hot soapstone can be used to continue the heat if the body becomes chilled before the sheet gets dry. This treatment removes very rapidly all chills, cures hay fevers, and produces a healthy action of the liver, where all other remedies have been sad failures. For any trouble in the water passage, or stoppage of the water caused by any inaction there, take a sitz-bath, remaining in the water hot as possible without burning, ten or fifteen minutes, rubbing directly over the parts affected with the hands, dipped in hot water. On retiring, take a light cloth, about two thicknesses, fifteen or twenty inches wide, and twenty-four inches in length, and fasten to the underclothing next to the skin, letting it rest directly over the bladder, and down between the limbs, letting it dry there during the night. Use soapstone or hot-water rubber bag if you are chilly, and drink plenty of hot water. If a few nights' treatment does not restore you, don't let medical doctors experiment on you by giving you spirits of niter or gin, which will soon ruin the water channel; but procure a good silver catheter, and draw off the water yourself. I have known persons to enjoy good health using this instrument for over thirty years for that purpose without experiencing any other inconvenience, and when this treatment doesn't restore the passage to a natural action, there is not any remedy that ever will. Get this silver instrument and use it; save your health, strength, and money by not allowing yourself to be experimented upon by medical doctors, for they will always be failures, as they ever have been. This will not, however, prevent you from seeking the advice of a good competent surgeon, who may possibly be able to relieve you by a surgical operation.

In kidney difficulties drink a cup of hot water before each meal. Rub the back and around the entire body on retiring with the hands dipped in hot water. Take a cotton band, two thicknesses, twenty-four inches wide, wring out of hot water, and pin it around the waist, letting it dry through the night, or you can wear it through the day if you wish, wringing it out of hot water as often as the heat of the body dries it. The same treatment for crick in the back or any form of lame back. Ladies will always find a hot water band to be their best friend for womb troubles, female weaknesses, and kidney troubles, using about one pint of hot water for injection once a day for any unhealthy discharge in water passage. Drink a cup of hot water before each meal. This is the safest and best treatment ever yet found. Don't fool your time and money away experimenting with medical doctors, who will always leave you worse than they found you. Be a friend to yourself.

Midwives will again find in child-birth these hot water bands or hot water cloths to be their most valuable aids. Renew them often, and give a few swallows of hot water to drink every few minutes. Hot water here, as in all other departments, has been and ever will be a most valuable aid in this direction. The true and only friend of mankind always sure and safe in its action when applied as I here direct; and this same treatment can always be used with success with all domestic animals, increasing or diminishing the dose according to the size or weight of the animal, using soapstones, hot bricks, or hot-water bags where heat is lacking or the animal appears to be chilly. Giving hot water inwardly, as you would any other mixture, is far better than all other mixtures ever heard of. Horses have repeatedly been cured by giving them hot water to drink. Give the horse warm water only until he is forced to drink it from sheer thirst.

I will also say a few words in regard to

keeping dogs and cats. The latter are not exterminators of rats and mice as is usually claimed. The rat and mouse will not usually live where there is no cat, and no well-regulated cat was ever known to even try to exterminate them. They usually eat one now and then, as you would a chicken, but are sure to keep a good stock on hand, and only when a cat gets insane does he ever endeavor to exterminate rats or mice. A few good traps are far better and less trouble than the filthy, dirty cat, as when not in use they can hang on a nail, and do not consume any food. And, again, when people become properly civilized, they will never consent to live in the same apartments with cats or dogs; and if the laws of our country allow people to keep them at all, they should be strictly confined in iron cages the same as other wild animals, and should always be killed when found running at large, and the state or town should pay a bounty for every head of a dog or cat when thus killed. The reader will doubtless be surprised at this radical statement, but I will give my reason why. Both the bodies of dogs and cats make a harbor for many dangerous parasites and insects, which are destructive and dangerous to human life. Many an innocent child loses its life, and many a person suffers disease for years, without being able to determine where this disease came from or what was its, cause. Worms inwardly, many of them,

are germinated from the cat and dog parasites, causing ringworm. Carbuncles and
boils have been traced directly to the cat
and dog, and that deadly tapeworm, so destructive to vegetable-eating animals and to
mankind, has been traced directly to the
dog, to say nothing about the victims bitten
by mad dogs. Where the dog ever saves one
life he destroys a hundred; where he saves
the owner one step he makes a hundred, and
where he saves his owner one dollar he costs
a hundred; and, as far as his usefulness in
hunting game is concerned, a few good guns
and traps are far better. We often meet in
the walks of life poor demented, miserable
men and women whose intellects are so de-
praved that they manifest an insane love for
filthy cats and vile, dirty dogs, instead of
lavishing that love that should be held so
sacred upon the members of their own families and the poor suffering children of men.
Strange fancy! Full developed insanity in
that direction, at least, to love a filthy dog.
Funny educators of children, indeed! Watch
the antics of a few male dogs as they meet
each other in the presence of refined ladies
and children on the public streets. Who is
there witnessing the daily scenes who is not
disgusted with a public sentiment which
enables nearly every family to keep from one
to fifty of these vile brutes to the annoyance
and disgust of every sensible man, woman,
and child? Again, listen in dark hours of

night, and hear the wailing of the miserable, dirty cat, disturbing alike the poor sick invalid and the weary workman and traveler who seek a few honest hours sleep so justly theirs. Listen again to the howling and barking of the miserable dogs far and near. In the morning they arise from their beds tired, disgusted, weary, and heart-sick at such a fearful state of society that permits and takes an active part in these disgusting affairs produced from or by the disgusting dog and cat. But the time will doubtless come when any one so demoralized as to be desirous of keeping that filthy varmint the cat and the vile brute the dog will be permitted only to do so by confining them in a good, substantial iron cage, never permitting them to run at large, under penalty of their being shot at sight or killed by the first person capturing them.

Kind reader, excuse this digression, but ever bear in mind that when mankind arrives to a proper state of civilization, the vile dog, and that filthy varmint, the cat, will no longer be allowed to run at large or occupy the same dwellings with the children of men ; and any man or woman who is so low in the scale of intelligence in that direction—at least, who is not able to find a more worthy and intellectual companion than a filthy dog or vile, filthy cat—is certainly in a very low condition in the walks of life ; but as humanity progresses, the medical doctor and the vile, filthy dog and cat must go, as they are, one and all, dangerous to the best interests of the children of men. Again, I would say in regard to health, that the care of the foot is very essential. The foot should be washed on retiring by dipping the hand lightly in hot water, and rubbing thoroughly between the toes and the hollow of the foot, and also all over, for that matter, as far up as the knee, at least every other night, as this treatment is a sure preventive of sweating feet, and no person giving the feet this treatment will long be troubled with cold feet or corns and bunions, as this treatment prevents any of these troubles, and is also a preventive of chilblains. Take care of your feet if you desire good health. By soaking the feet in hot water some people take cold readily; therefore rub them as here directed, and you avoid this.

For raising of blood or bleeding of lungs or stomach, take one-half teacupful of hot water every half hour; will give immediate relief.

Any person who is so unfortunate as to have an extreme desire for intoxicating drinks can easily subdue that extreme thirst by drinking hot water at the rate of one teacupful three times a day. Any person who has been so unfortunate as to become an opium eater, and desires to break off this vile, filthy habit, can find a perfect cure by taking a hot-water bath on retiring, with water hot as possible

without burning, and drinking three teacup-
fuls a day—one before each meal, hot as possible without burning. Eighteen or twenty
days of this treatment cures the worst forms
of opium-taking. And morphine-takers can
cure themselves by this same treatment. In
regard to any special degree of heat, no two
persons can bear the same degree of heat,
but everyone should always use it hot as possible without burning. Every family ought
to have a bathtub in the house, if it is nothing more than a common sitz-bathtub. Let
any person drinking one cupful of hot water
each morning before eating take a hand-bath
at least one night in the week before retiring, as a sure preventive of biliousness
and fevers, and a preventive of ever catching any contagious disease, as smallpox,
cholera, yellow fever, or any of these filthy
diseases. Again I caution you against vaccination, one of the greatest curses of this
enlightened age—an immense scheme of medical frauds, called doctors, who are ever on
the alert to fill the bodies of men, women,
and poor, innocent children with the germs
of disease, which prepare their bodies in
such a manner that they are victims of medical doctors during the remainder of their
natural lives, as thousands of their victims
can testify whose bodies have been ruined
by vaccination. Therefore be wise, and never
allow yourselves or families to be vaccinated
for any disease; and ever bear in mind that

this so-called medical science has ever been a failure, and ever will be, and that cures by electricity, too, are like doses of medicine— only temporary reliefs. Electric garments and all electric fixtures are only a relief for the time being, like faith and mind cures, which do not produce any radical change in the body; that hot water used as here described is the only renovator and true cure ever yet discovered by mortal man, and that its use in public hospitals and in all public institutions will in a very few years destroy the use of medicine, and do away with those curses of humanity, patent-medicine-makers, medicine-sellers, medical doctors, and drug stores. Those great curses to mankind must be destroyed forever, and any person who thinks that medical doctors give people medicine to cure them of disease must be very silly indeed. Nearly all diseases are made worse by the use of medicine. If the sick are helped in one direction, they are sure to be injured in another, and thousands of victims are burnt up, nearly, and suffer for twenty days or more with fevers that might easily be broken up entirely in twenty-four to forty-eight hours, as the medicine given by doctors is usually given to feed and prolong the fever, instead of breaking it up, as these doctors try to make you believe.

Very fleshy persons who desire to reduce their flesh will use hot water the same as for consumption. The use of hot water for the cure of all diseases was practiced by the ancients for two thousand years, and medical doctors were unknown. The art or method of using it was known and practiced by the common people, and this must have been a very happy age of the world when its inhabitants were not cursed with doctors or heavy doctors' bills, and two-dollar prescriptions to drug stores for some useless medicine worth about eight cents, first cost. The use of hot water as a medicine proves to be one of the lost arts again restored to bless the human family. It was used three hundred years ago with great success by noted doctors in Germany, and is used very extensively in curing all forms of disease in many parts of China, and by many other Eastern nations at the present day for the same purpose. Some people are very partial to putting something in the water, thinking to improve it, but nature abhors a compound unless it is compounded by herself. Always use the water clear. Some invalids go all the way to the Hot Springs, to Arkansas and other parts of the world, and fraudulent doctors and pretended scientists, and other interested persons strive to make people believe that it is some mineral or other ingredient contained in the water that performs the cure. That is another deception. It is not any substance in the water that performs the cure, but it is the volume of heat in the water that acts independent of any substance contained in the water. The heat performs the cure, and hot water from the teakettle at home will do just as well as hot water in the Hot Springs, as it is the heat that is wanted to perform the desired work. Many people go to Saratoga and other springs to drink spring water for their health. It is the volume of water that improves the health, and not the mineral which the water contains, as they usually suppose ; and the same amount of pure water drank at home would do as well; if drank hot would do much better. Other people go to the seashore for salt-water baths, but it needs to be a very strong person to stand these salt-water baths, and only very healthy people should take them. To sickly and very weakly persons they are decidedly an injury. A hand bath of hot water applied by themselves or applied by some strong, healthy person on retiring at night would be far more beneficial, and clear water at that, instead of salt water, is far better. Then be wise. Use hot water as I here direct, and prove the truth of my assertions. You can use this great curative agent at home, and be blessed by curing yourself and your friends and neighbors.

Yours in love, friendship, and truth,

THE AUTHOR.

GENERAL RULES TO BE OBSERVED
IN THE USE OF HOT WATER.

———o———

Very many persons, not satisfied with using a moderate amount of hot water, drink it with their meals, but one teacupful or half-pint is a plenty before each meal. Three times a day for rheumatic troubles, in connection with other treatment; rules laid down on other pages of this book. Most people troubled with this complaint are very fond of food highly seasoned with salt. While they need not quit it entirely, it should be used very lightly, as its excessive use is a great producer of rheumatic troubles. Again, all families should have one stated rule for the use of hot water. Let each member over seven years of age drink immediately before eating in the morning one half-pint of hot water. Always drink hot as possible without burning. And make one stated rule. On changing the underclothes—usually Saturday night, the most convenient—take a hand-bath, dipping the hands in hot water and rubbing the entire body all over on retiring, chafing the body dry with a cloth or towel. All members of the family under seven years, drink one-half teacupful. Follow these rules and rest assured that no member of the family will ever have a fever, or catch any contagious disease, or be troubled with biliousness, as this treatment prevents all these, chills and fever as well as the rest. For sprains or any swelling of the knees or ankles rub the parts affected with the hands dipped in hot water; bind hot bands around them wrung out of hot water, covering with dry bands; renewing this treatment as often as the heat dries them. Hot water is a powerful tonic for the stomach when 'drunk in small doses as here ordered, but when drunk in large doses becomes weakening to the stomach. Therefore use it with care, as here directed.

For catarrh of the bladder, or any irritation of the water passage of any name or nature, take sitz-bath on retiring, water hot as possible without burning, remaining in the water ten minutes, and injections of hot water in water passage, using a pint or more to cleanse the water passage faithfully, using the water hot as possible without burning. Also on retiring place a hot cloth wrung out of hot water across the lower part of the bowels and between the limbs and directly over the bladder, letting the heat of the body dry it through the night. This has been found to cure this trouble quicker and better than any remedy ever yet found by any medical expert that ever walked this earth.

Piles and all gatherings and sores in the back passage can be relieved and cured with sitz-baths and hot water injections, taken usually on retiring at night. The very worst gatherings and most troublesome sores and humors connected with the back passage are readily cured by faithful treatments of hot water used on retiring at night. At the same time a teacupful must be drunk immediately before each meal three times a day. At the same time take the weekly hand-bath one night in the week on retiring, as herein directed. Treat all formations like tumors or cancers on any part of the body as here instructed. In many instances these unhealthy formations can be eradicated entirely by this treatment, and any formation or tumor or cancerous growth can be quieted and kept in subjection by the use of hot water, hot compresses, by drinking, and treating the entire body once or twice a week with hand-baths as directed in this book. Treat venereal diseases and all sores and troubles of this nature on all parts of the body with hot water as here directed, and they can readily be cured where they are, without being scattered or driven to other parts of the body, as they usually are by taking vile, poisonous medicines usually given by medical doctors for these troubles. Again, treat the feet at least twice or three times a week with your hands dipped in hot water on retiring at night, and you will avoid chilblains, corns, and bunions, and if you already have them, treat them as here directed and you will readily remove them without pain or having your feet butchered by any modern corn doctor, and save time, pain, and money by not buying any of those miserable preparations usually sold at drug-stores for curing corns and bunions—miserable failures, like all other preparations sold at these places. Again, my friends, don't buy any of their miserable, vile medical preparations, for they are to-day as miserable and sad failures as they ever have been in every age of the world and as they ever will be in all time to come. Remember again, kind reader, wherever you find any community in any part of the world who are not intellectual enough to regulate their general health without resorting to drug-stores for the vile preparations sold there, you may rest assured that such communities are in a very low state of intelligence, at least in that direction. We frequently see apparently intelligent gentlemen with broadcloth coats, and fine-appearing ladies with sealskin sacques, who frequent these gilded dens of vice and buy these vile preparations sold there as medicine, yet they look with disdain upon the common workman and tramp who buy vile preparations to recuperate their exhausted energies at the bar of some low gin-mill; but let me modestly tell you, my friend, whilst I do not advise you to patronize either, those who buy at the gin-mill get a better quality of drink, less poisonous, less injurious, than those who patronize those gilded dens of vice, the drug-stores, notwithstanding their apparently respectable airs and appearance. Ever

bear this straightforward, truthful statement in mind, for truth will ever bear
repeating. Do not get frightened by
the statement of any apparently learned
ignoramus of a medical doctor who gives
your disease some outlandish name, hoping
thereby to scare you by making you think it
is a new disease and more dangerous than of
old, for disease in any part of your body ever
has been and ever will be the same, taking
different forms with different persons according to their organization. We can truly say
that disease is a unit, and hot water a universal cure for all diseases of man or beast,
when properly and faithfully applied.

In regard to using cold water, it will not
act until hot, whether it is heated by your
body, or used hot, when it will act immediately. Any person naturally strong and
healthy can get a good result and find it very
beneficial to their general health. On rising
in the morning dip the hands lightly in cold
water, rub the entire body briskly all over,
then chafe dry with cloth or towel. This
produces a very fine, exhilarating sensation
all over the body, lasting through the day,
and is very good for the general health, but
only strong and healthy persons should
practice this. Heaves in horses can readily
be cured by giving hot water to them to
drink, beginning with it slightly warm, gradually increasing the heat day by day until
you get it hot as can be drunk without burn·

ing. This is the first and only remedy ever discovered that cures the heaves in horses, which is nearly the same as asthma in people.

Small children should be given daily a few teaspoonfuls of hot water to keep them in general good health. In case of a cold or sore throat or being threatened with fever, when you are placed where hot water cannot be readily obtained, a person can get a good result and break the trouble by drinking one great-spoonful of cold water once in twenty minutes, as the body will heat it and the desired result can be obtained. Persons about to go out and be exposed to severe cold weather should drink a cup of hot water immediately before going out, and any person having been out and got chilly or wet should drink two or three cups on coming in, as that will warm up the body in a few minutes and prevent taking cold. It is the safest drink for all diseases named in this book in the entire universe.

Persons having diarrhœa or piles, which are generally or often produced by taking pills or physic, should at once drop the use of meat as food; let vegetables, bread, and fruit take its place. As meat of any kind is heating, and makes any of these troubles much worse. For cuts, wounds, bruises, do up the injured part with stout, new cloth two or three thicknesses and wet with hot water as often as the heat of the body dries it. This will heal up a severe cut, wound, or bruise quicker than any other remedy in the known world, and allay all inflammation. If you wet with cold water, the heat of the injured part must heat it before it acts, but when used hot it acts immediately.

To improve the voice for public speaking and singing, rub the breast and throat on retiring at night with the hands dipped in hot water; drink two or three great-spoonfuls the last thing as you retire. If you were where hot water could not be readily obtained, cold water could be used in the same manner, but more rubbing would be necessary to get up the heat. Also drink two or three great-spoonfuls of water immediately on rising in the morning. This is a good treatment for all throat troubles and bronchial difficulties, and therefore have courage.

Always treat all diseases with hot water, using it faithfully, and if any remedy can possibly save you, this will. And in case you lose a member of your family, and this treatment has been faithfully applied, you may rest assured that all that could possibly have been done by human agency has been done. And if anyone disputes the statement that medical doctors are a curse to humanity and kill three where they save one life, let them visit the graveyards where their victims are deposited, or let them examine the death rates in cities and villages where those gilded dens of vice, the drug-stores, and medical doctors are the most plentiful. And ever bear in mind this one important fact, that any doctor who does not have knowledge enough to treat diseased persons without medicine is a confoundedly ignorant individual. THE AUTHOR.

———o———

MINUTE INSTRUCTIONS TO BE OBSERVED IN THE GENERAL USE OF HOT WATER.

No person in ordinary health who will drink one half-pint of hot water before eating in the morning will ever become bilious, or be troubled with a sore throat, or be subject to any form of fever, or have any attack of that dread disease called la grippe, or be subject to any form of the cholera. Hot water should always be drunk hot as possible without burning the mouth or throat. It should be used in the following manner : Take a half-pint teacup with a saucer, pour it out hot, as people pour out hot tea, blow it slightly, and sip it as you would drink hot tea, occupying about five minutes in drinking. A half-pint used as above described as a preventive of disease, as well as a cure. If drinking hot water makes you sick, with a tendency to vomit, it shows a very bilious condition of the stomach, and hence the greater need of using it as herein directed. If it causes you to vomit, don't let that scare you, for even

then it will do you good, as vomiting by its use
will throw up the bilious matter and leave you
in a more healthy condition. This vomiting
does not usually take place more than from one
to four days. In case of severe or acute pain or
cramps, where hypodermic injections of morphine have formerly been used, skin injections
of hot water have been found far superior, as it
gives immediate relief when applied in the same
manner, and does not leave that bad condition
that is usually caused by hypodermic injections
of that vile drug known as morphine. Any vic-
tim who has been demoralized by its use will
find sure and permanent relief by injecting hot
water in its place, using it in the same manner
as morphine has been or is used for skin injections. Any person who has been accustomed to
use morphine will know how to apply hot water
in its place. In regard to the best treatment
for cholera ever yet found, treat a cholera patient in the following manner : Inject about
two quarts or more up the back passage ; have
it as hot as possible without burning, adding
one great spoonful of fine salt. Also give
hypodermic injections of the same directly over
the bowels, using a quart or more in the space
of three hours. Continue to apply it in the
above manner until the person is fully relieved.
Give the person one cupful or one-half pint of
clear hot water to drink every half-hour until
relieved. This proves to be one of the best
treatments for this dread disease ever yet discovered.

To break up a severe attack of la grippe the following treatment has been found the best of any yet tried: Give one-half pint of hot water to drink every half-hour; wring out a sheet from hot water and wrap it around the entire form, covering with woolen blankets or comforters, and using soapstone or hot water rubber bag to keep from chills. Renew the hot sheet as often as the heat of the body dries it. Drink three or four cups of hot water for ten or fifteen days after getting relief, which usually takes place after from twelve to forty-eight hours' treatment. While using the hot sheet to break up the severe pain drink a cup of hot water every half-hour.

This same treatment is very good for severe rheumatic troubles, as well as la grippe. This is one of the best treatments ever yet found for all rheumatic troubles. It has also been proved that hot water used as I here direct is one of the best treatments yet discovered for kidney trouble, diabetes, liver troubles and constipation, and even consumptives have kept themselves alive for many long years by using hot water as follows: Once in five to fifteen days; in very severe troubles, for a few days use it every other day for ten or twelve days, gradually slacking off to a treatment once in from five to fifteen days. Great good has almost universally been accomplished by using hot water as follows: Using from two to four quarts by injecting it into the back passage on retiring at night, that being the most conven-

ient time. Inject four quarts, if possible, or as
much as you can, at least, by lying on your
back, with a small pillow under the small of
the back, letting the back be slightly elevated.
A much larger amount can be injected than any
would naturally suppose. Let this remain
until you have a thorough operation. This
large amount cleans out the rectum and large
intestine, where many germs of disease often remain and disease the entire body. When this
has operated throw up one pint of clear hot
water, and lie still in bed, retaining this last injection during the night. This, being used in
this manner, will remain and pass out during
the night or the next morning through the front
passage, which is one of the most beneficial
remedies for the liver and kidneys ever yet
found. At the same time drink three half pints
of hot water, or one before each meal, during
the day. For consumption or weak lungs, drink
three pints a day, one before each meal ; also
eat stale bread and beefsteak prepared in the
following manner : Chop it fine, as you would
mince-meat, make up into small balls, as you
would make fish balls, have your fry-kettle or
griddle hot, and brown them over on the outside
but have them rare done, and eat them hot. Let
them eat all the ripe grapes, oranges, pine-
apples or pine-apple sauce they wish ; also let
them use good sweet cream, with loaf sugar.
This same treatment has been used with great
success to reduce flesh. At the same time let
them reduce their rations too, or not eat too

much or too often, as we have many persons in this country who dig their graves with their teeth, or rather prepare their bodies for the grave with their teeth, bringing on many diseases by over-eating. Therefore, follow strictly these instructions within this work and you will secure long life and be happy, by using the only safe universal remedy ever discovered by mortal man for curing all forms of disease that humanity is subject to. Again, are you troubled with a lack of sleep, or rather wakefulness, on retiring, caused by overtaxing the body in any direction? Just before retiring drink from two to three cups of hot water. This will enable you to have a good night's rest. Again, always have the water you use to drink boiled—letting it boil fifteen or twenty minutes—as boiling destroys all impure germs in the water, and renders it perfectly pure. Let it cool before drinking, and ice it if you wish before drinking. Remember, again, that in all workshops and public drinking-places hot water should be used for drinking purposes always, in place of cold water, as it quenches thirst much quicker, leaves one in a healthy condition, where drinking cold water often inflames the stomach and intestines, causing dysentery, but drinking hot water prevents all this, leaving the stomach in a perfectly healthy condition; therefore drink hot water and be healthy and happy.

Hot water used in the following manner cures all ear troubles, deafness, severe pains, noises,

or unpleasant sounds—often restores the hearing. Take some hot water in a cup, temper it
so it will not burn the inside of the ear, but
use it as hot as you can without burning;
turn your head slightly to one side; turn in two
teaspoonfuls, hold the head in this position
about two minutes, rubbing the neck lightly
directly under the ear with the fingers, then
straighten up, let the water run down the
neck and rub it in faithfully with the hand
as you would rub in a liniment of any
kind. Treat both ears in same manner. This
treatment once a day has often restored the
hearing where people have been deaf for
many years. Also treat all gatherings or running sores in the ears in the same manner.
For gatherings or severe pains in the ears
apply hot water in the same manner, applying it as often as the severe pain makes it
necessary until relief is obtained; also place a
hot cloth over the outside of the ear, covering with a dry one, renewing it as often as the
heat dries it. Can also wet a small piece of
wool or cotton and place inside the ear.
    Wetting it with hot water—this is the finest
treatment ever yet found for any trouble of
the ears. Also treat all boils or sores, swell-
ings, formations on the body or limbs, of any
name or nature, in the following manner:
Place a thin cloth over the affected part,
let it reach down into a pail or vessel to
catch the water, and pour the water on lightly
for fifteen or twenty minutes, twice or three

times every twenty-four hours. This treatment gives relief to all troubles of this kind where all other remedies fail. Treat sprains and bruises in the same manner; use hot cloths the rest of the time until relief is obtained. This removes all discolor and removes pain and soreness as well. All small children should be trained to drink a small cup of hot water at least once a day, and all infants should be given three or four teaspoonfuls of hot water once or twice a day, as that would be a sure preventive of sore throats, croup or diphtheria.

For pimples on the face the following treatment has been found very good in connection with other treatment: In this work all persons should always wash the face in hot water, as it preserves the eyesight, at the same time makes a fine, smooth countenance, keeps the features smooth and free from wrinkles. With young or old, for sores or pimples on the head, face or neck take a sponge or piece of cloth, dip it in water hot as possible without burning and hold it on them; renew as often as it gets cool. Give this treatment fifteen or twenty minutes each day. From your Friend and Brother,

THE AUTHOR.

www.ingramcontent.com/pod-product-compliance
Lightning Source LLC
Chambersburg PA
CBHW021957190326
41519CB00009B/1295